Wilderness Warrior Forest Adventures

The Math Workbook Kids Actually WANT to DO

ages 3 – 6

by Ginger McBride

Wilderness Warrior: Math Edition

The Math Workbook Kids Actually WANT to Do

Author: Ginger McBride

For permissions
Email: Ginger@GingertheTeacher.com
www.GingertheTeacher.com
www.grdmpublishing.net

Be sure to get your free worksheets that are updated every month.

How To Use This Book

Mission Board

Welcome to your wilderness math adventure! This book is full of missions, badges, and forest fun to help you learn all kinds of math skills - from counting and sorting to measuring and patterns.

YOUR MATH MISSION MAP:

- Start at the beginning and go one page at a time - or pick your favorite activity!
- Every section has a Wilderness Badge to earn, Just complete the fun tasks to unlock it!
- Use your hands, your eyes, and your brain - you'll draw, color, count, hop, sort, and even play!
- Some pages have a place to cut and paste - ask a grown-up to help you with scissors.
- Don't forget to check off each badge on the Badge Tracker when you finish a mission!

◎ Are you ready to explore, learn, and play in the wild?
Let's go, little
Math Explorer!

Wilderness Warm-Up:

How to Hold a Pencil

To help a child hold a pencil correctly using a hair tie, you can follow these steps: Wrap the hair tie around the child's thumb and index finger to guide them in holding the pencil properly.

This method encourages a correct grip by gently pulling the pencil into the web space of the hand, which helps develop fine motor skills.

This technique is effective for helping them form good writing habits. This is simple and can be a fun way to teach children how to hold a pencil correctly.

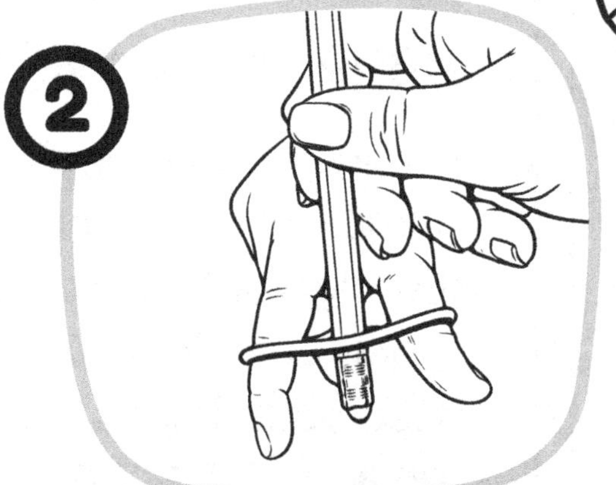

Explorer Tip: Hold your pencil gently – not too tight!

Grown-Up Note: This grip helps with control for circling, drawing shapes and writing numbers.

Math Mission Map

Tools You'll Need

Count The Animal Tracks

Look at each set of animal tracks. Count how many you see, then write the number in box!

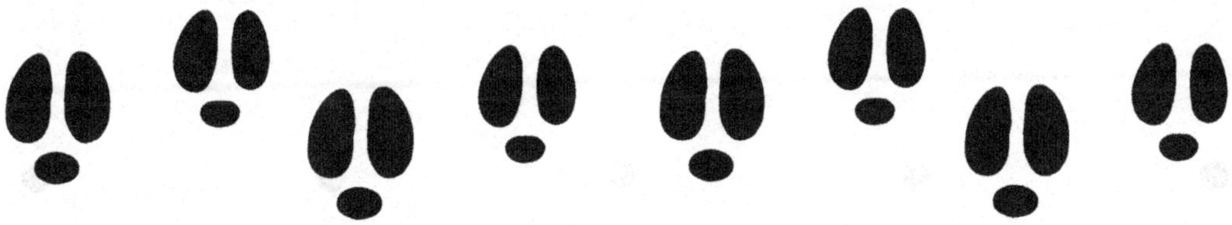

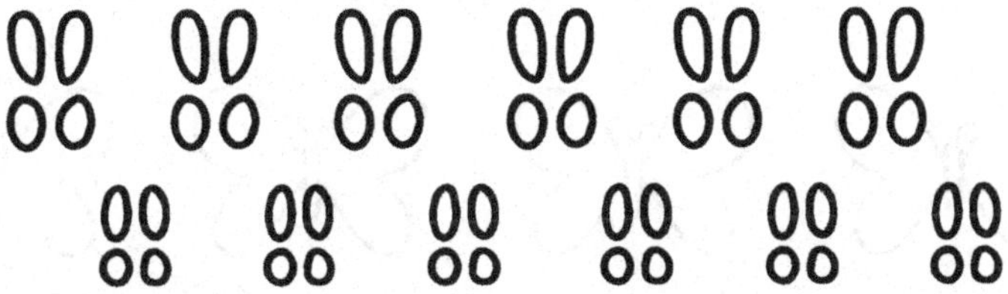

Tracing Numbers

Trace the strokes, numbers and words, then write on your own.

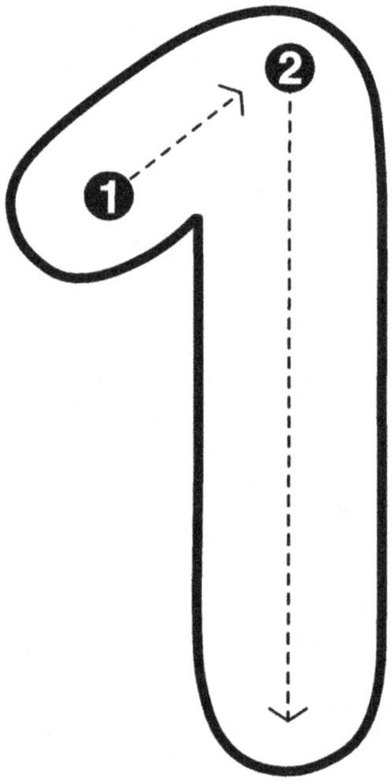

Count 1 butterfly and circle them!

Tracing Numbers

Trace the strokes, numbers and words, then write on your own.

Count 2 trees and circle them!

Tracing Numbers

Trace the strokes, numbers and words, then write on your own.

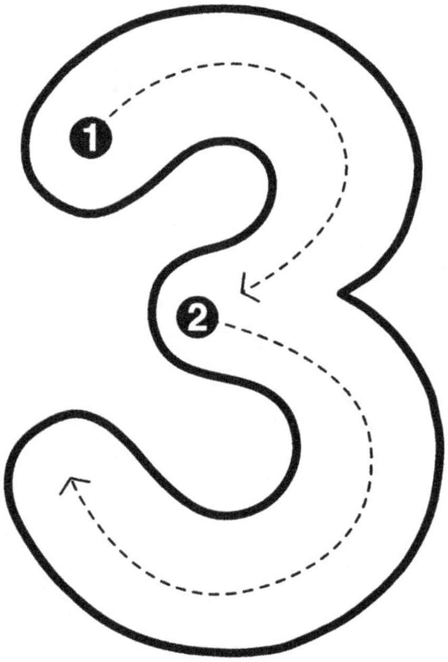

Count 3 leaves and circle them!

Tracing Numbers

Trace the strokes, numbers and words, then write on your own.

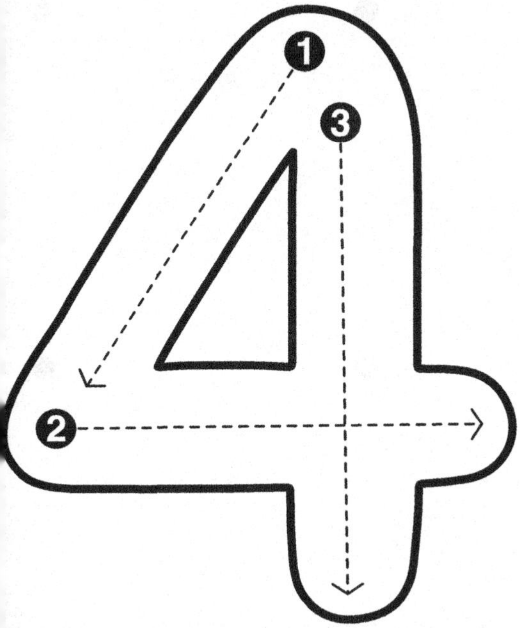

Count 4 features and circle them!

Tracing Numbers

Trace the strokes, numbers and words, then write on your own.

Count 5 flowers and circle them!

Tracing Numbers

Trace the strokes, numbers and words, then write on your own.

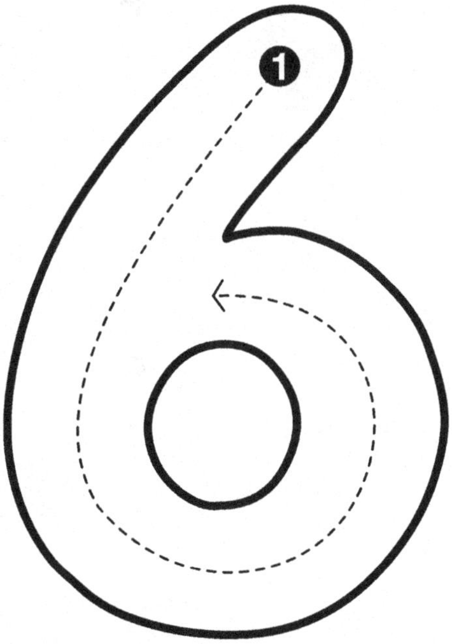

Count 6 stones and circle them!

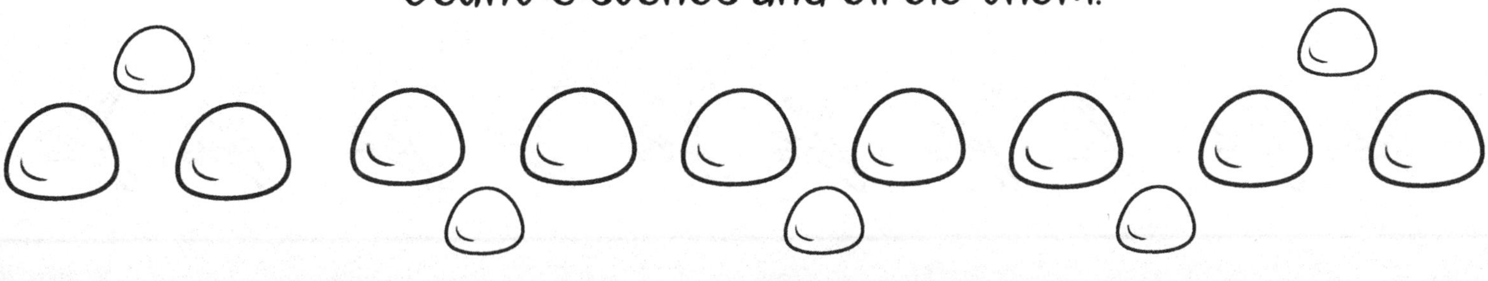

Tracing Numbers

Trace the strokes, numbers and words, then write on your own.

Count 7 sticks and circle them!

Tracing Numbers

Trace the strokes, numbers and words, then write on your own.

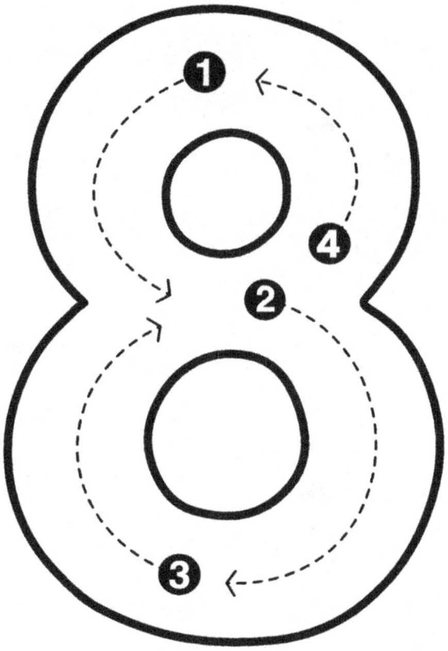

Count 8 berries and circle them!

Tracing Numbers

Trace the strokes, numbers and words, then write on your own.

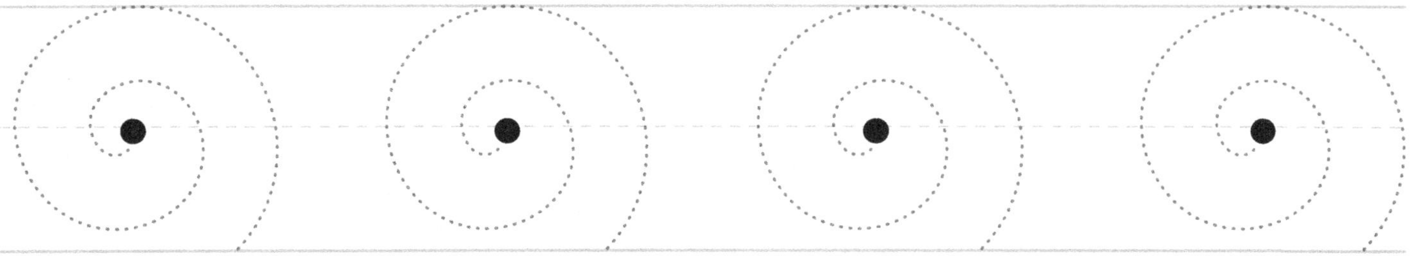

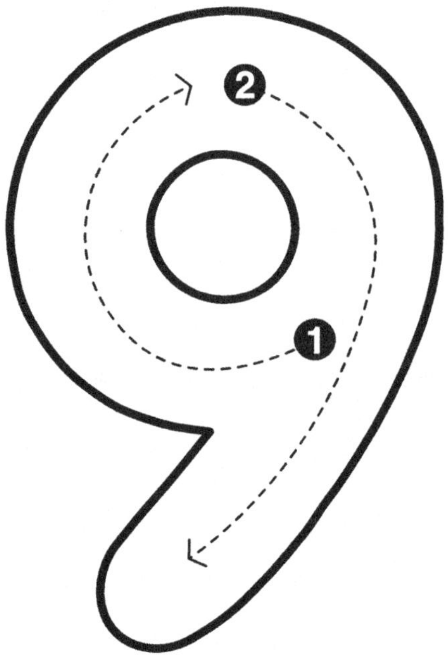

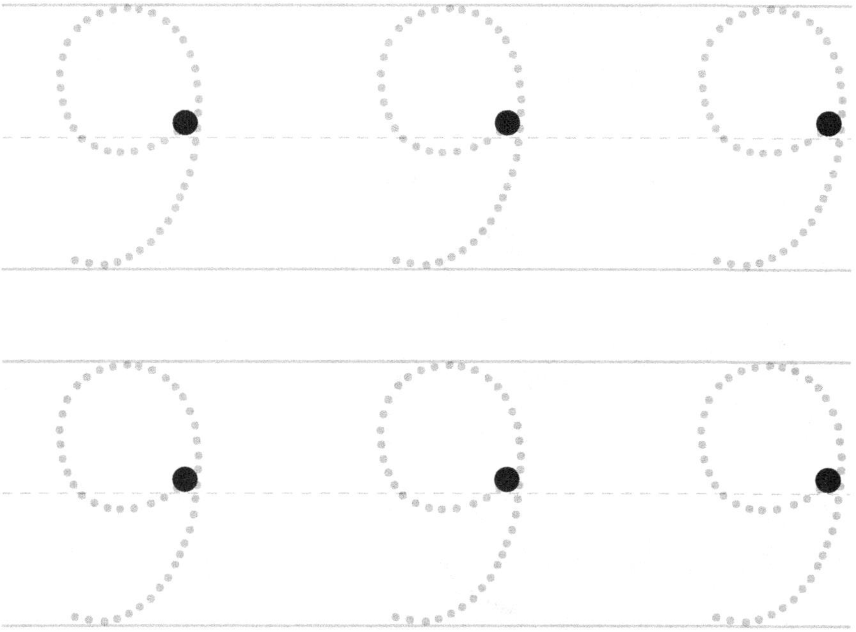

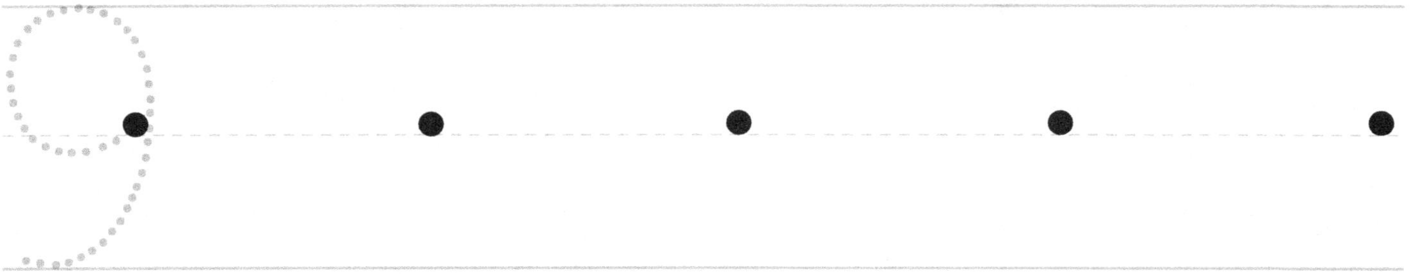

Count 9 bears and circle them!

Tracing Numbers

Trace the strokes, numbers and words, then write on your own.

Count 10 mushrooms and circle them!

Tracing Numbers

Trace the strokes, numbers and words, then write on your own.

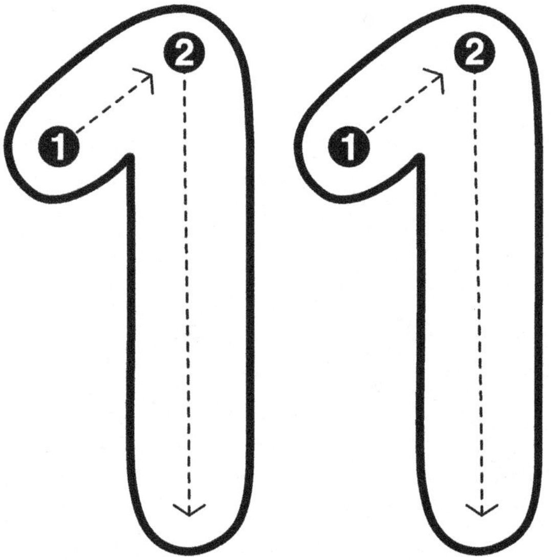

Count 11 nuts and circle them!

Tracing Numbers

Trace the strokes, numbers and words, then write on your own.

Count 12 mushrooms and circle them!

Forest Number Match

Draw a line to match each group of animals to the correct number.

3

4

2

5

1

Dot The Fireflies (1–12)

Use stickers or crayons to dot the same number of fireflies as the number on each jar!

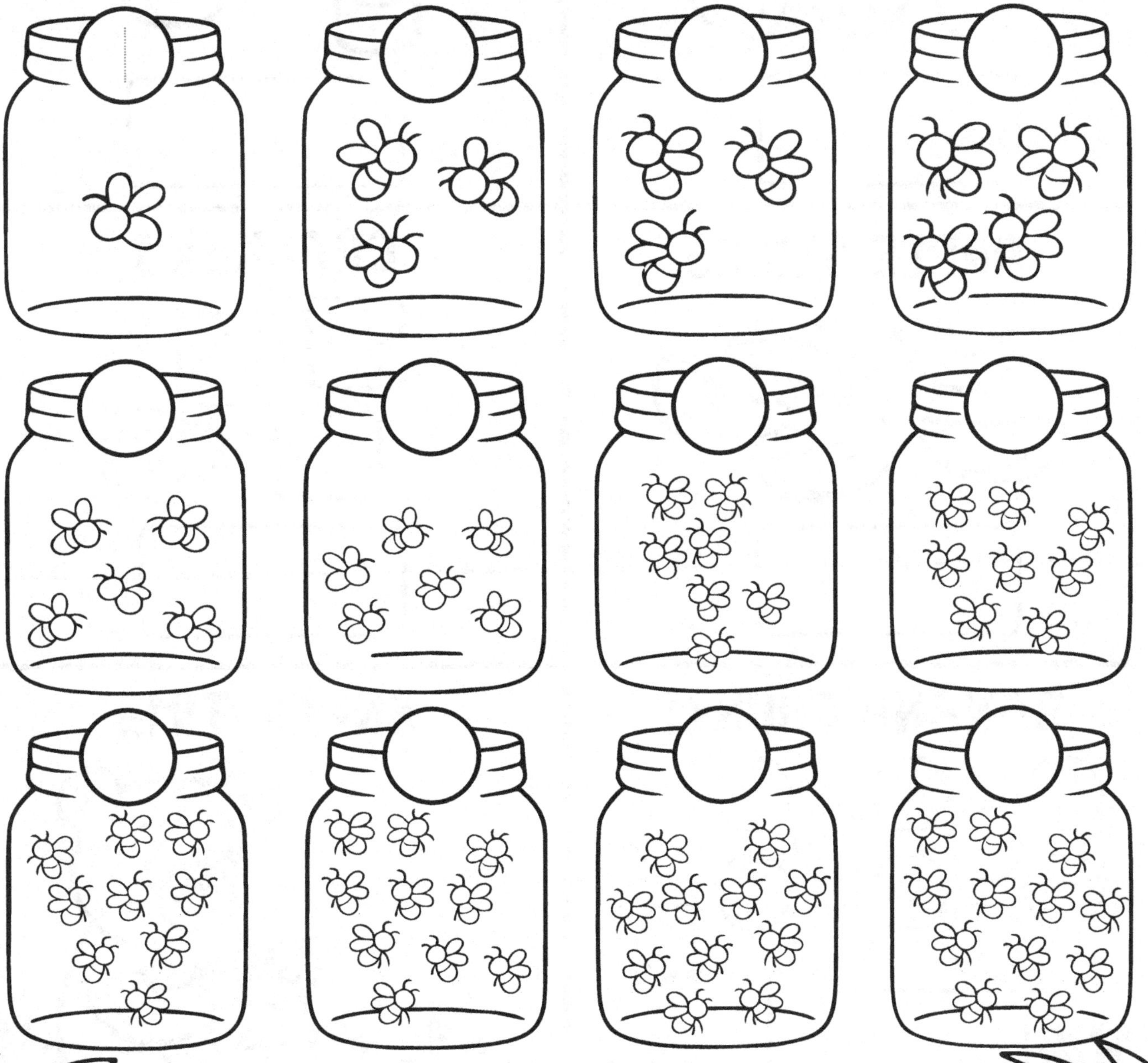

Snack Time Counting

Count how many snacks are in each group.
Then write the number in the box below.

BERRIES

MARSHMALLOWS

APPLE SLICES

CRACKERS

BANANA CHUNKS

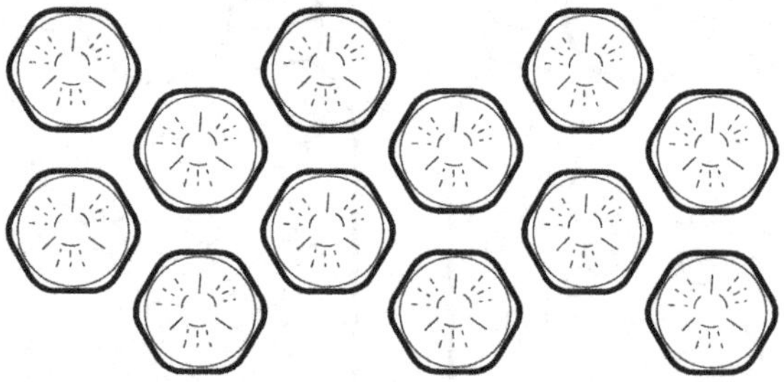

SNACK TIME

Number Line Hop Game

Hop along the number trail!
Say each number out loud as you land on it.

Number Hide & Seek

Can you find the numbers 1 to 10 hidden in the forest?
Circle each one!

Campfire Numbers Worksheet

Count each group of campfire items,
Then write the number in the box!

Sort The Forest Leaves

Look at each leaf. Write the correct number into the right box - how many are round, pointy, or long?

ROUND	POINTY	LONG

Toy Rescue Sorting Mission

Help sort the animal toys! Which ones fly, swim. or crawl? Write the correct number of each group into right zone.

FLY	SWIM	CRAWL

Button Sorting Worksheet

Sort the buttons! First by size, then by pattern.
Write the correct number in the right boxes.

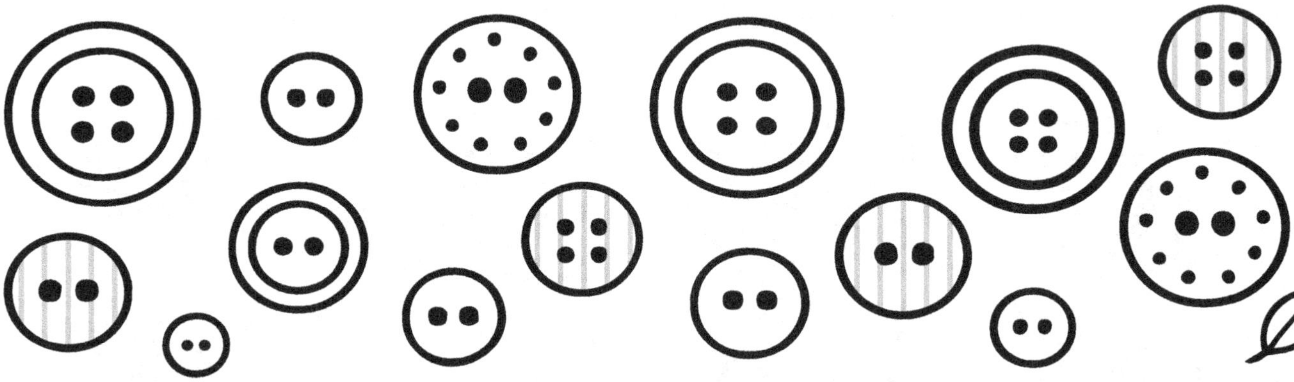

SORT BY SIZE	
Small	
Medium	
Big	

SORT BY PATTERN		
Dotted	Striped	Plain

Match The Socks!

Can you find the matching socks?
Draw a line to connect each pair!

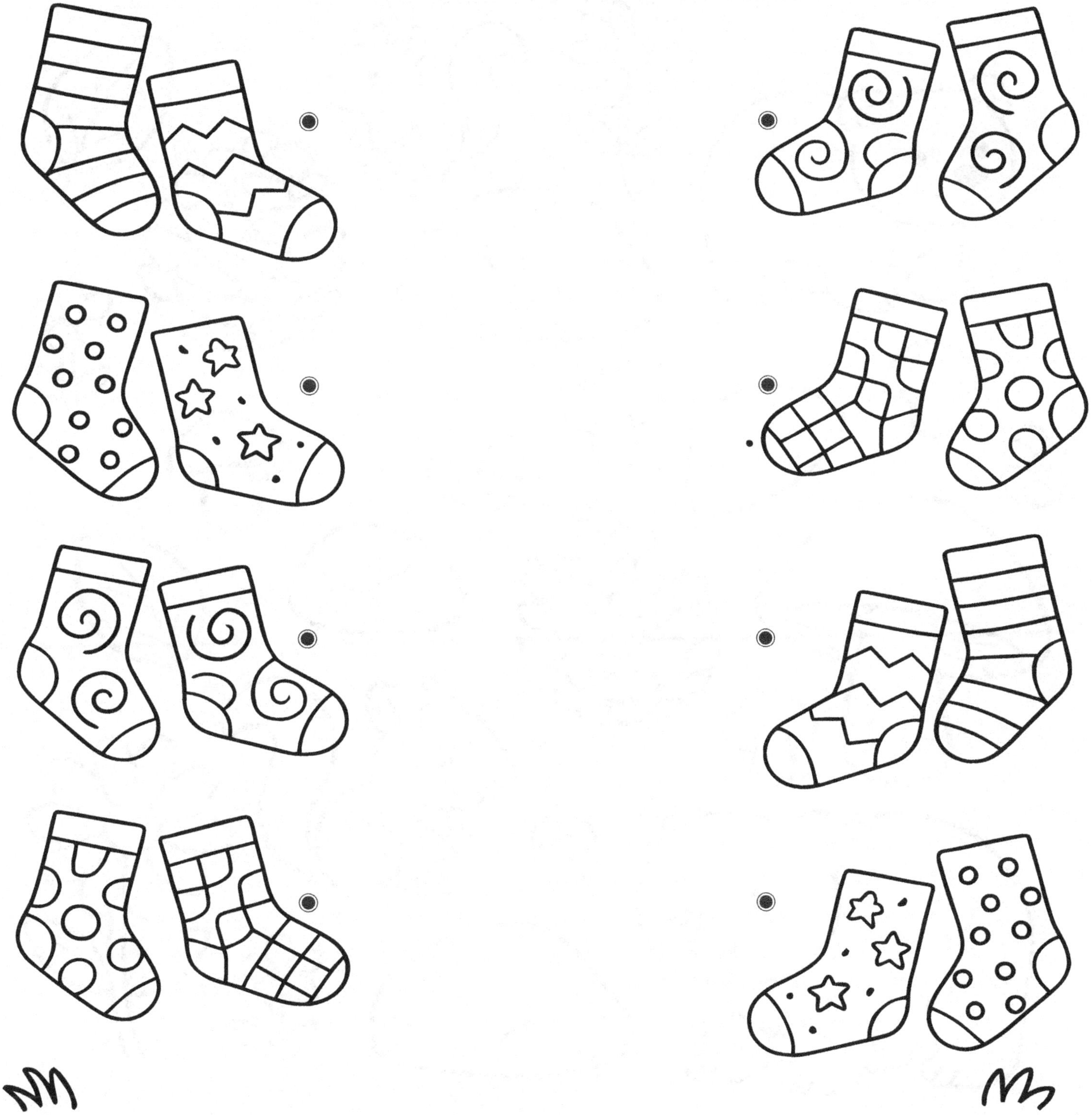

Sorting By Size
Big & Small In Nature

Sort the forest items by size.
Color the big objects using a dark color.
Color the small objects using a bright color.

Build A Shape Tent

Use the shapes below to build your own tent!
Cut and paste them into the picture.

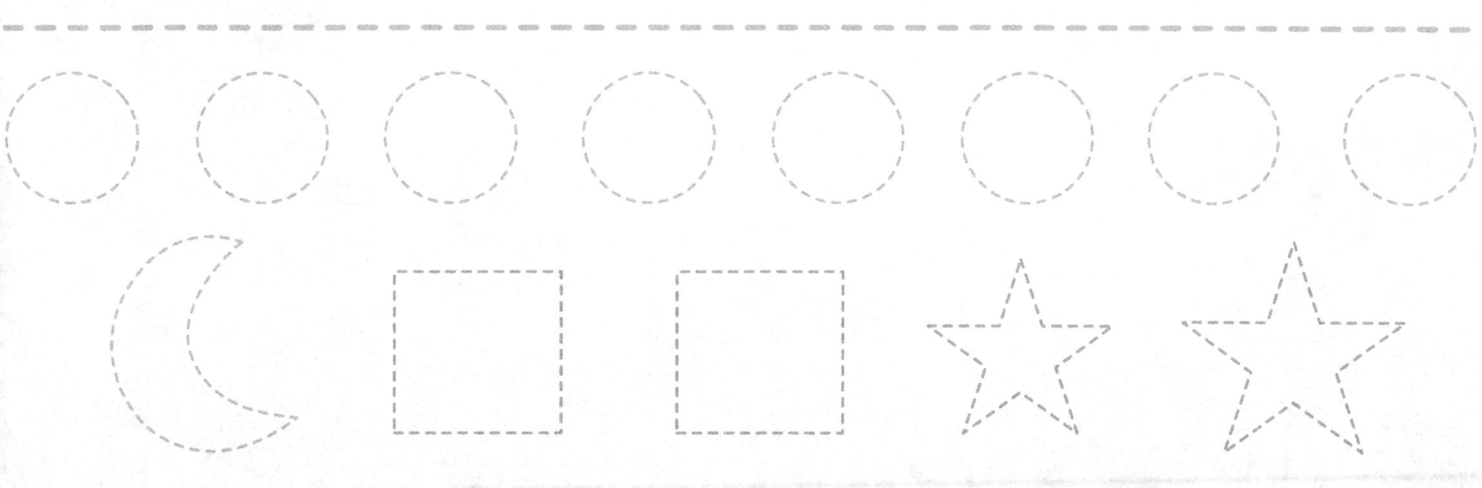

WILDERNESS WARRIOR CROSS BORDER

Obstacle Course Map

Trace the path from Start to Finish.
Follow the directions as you go!

Shape Scavenger Hunt

Can you find all the shapes hidden in the forest?
Color them in and check them off below!

- Circle
- Star
- Square
- Rectangle
- Triangle

Playdough Shape Mats

Trace, build, or color each shape using crayons!

Circle

Rectangle

Triangle

Square

Star

Inside, Outside, Beside – Forest Friends

Look at where each animal is.
Circle the correct word: INSIDE, OUTSIDE, or BESIDE.

Measure The Log!

Use the sticks to measure the log.
Count how many sticks long it is and write the number.

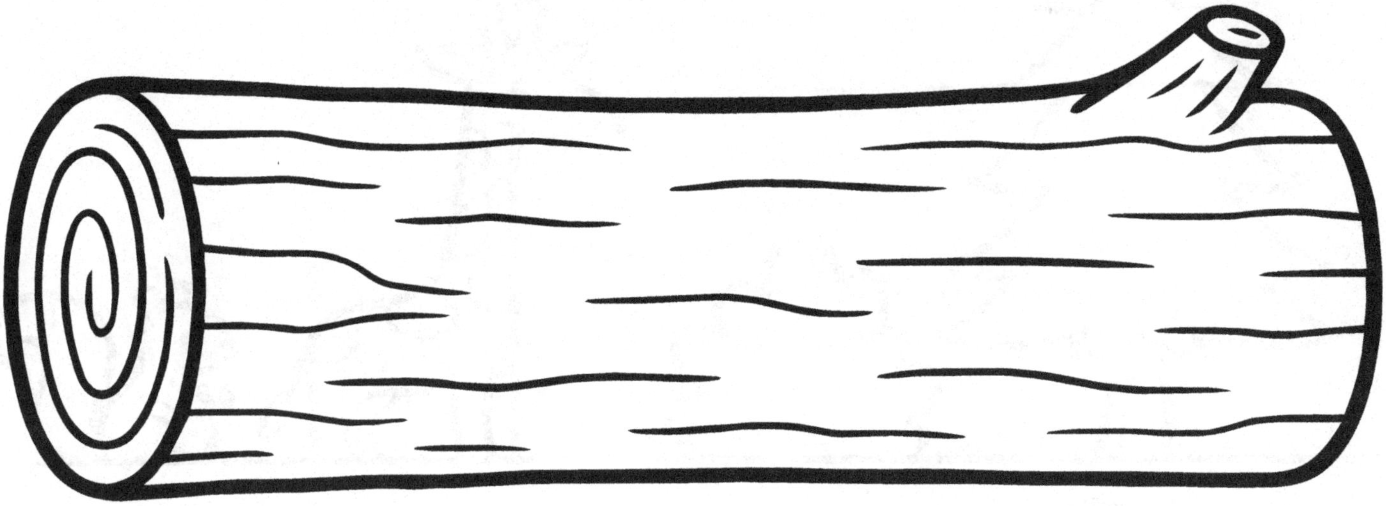

HOW MANY STICKS LONG?

1 2 3 4 5 6 7

Which Is Taller?

Look at each pair. Color the one that is taller.

Heavy Or Light? Sort The Gear

Look at the camping gear.
Decide which items are heavy and which are light.
Write the correct number of each group in the box below.

HEAVY	LIGHT

Paper Strip Races

Look at the race tracks. Which one is the longest?
Color the longest track in each race.

Water Cup Measuring

Look at each cup. How much water is in it?
Tick the correct word: FULL, HALF FULL, or EMPTY.

- ○ FULL
- ○ HALF FULL
- ○ EMPTY

- ○ FULL
- ○ HALF FULL
- ○ EMPTY

- ○ FULL
- ○ HALF FULL
- ○ EMPTY

Trail Mix Measurement

Count how many pieces of trail mix it takes to measure each item. Write the number in the box.

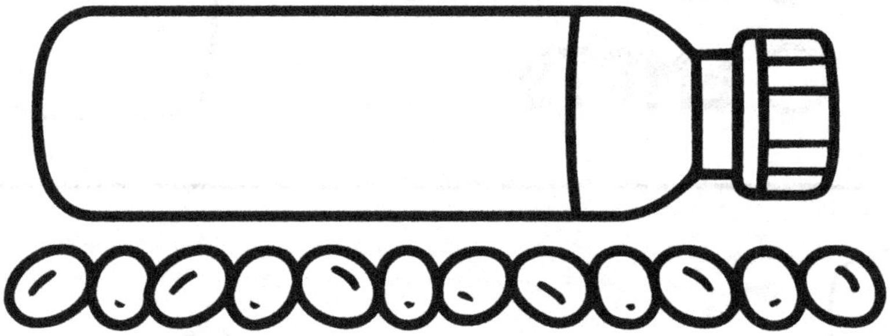

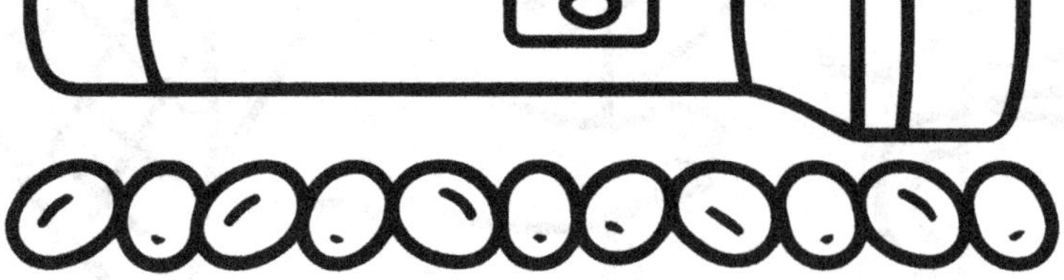

Compare Tent Sizes

Look at the tents.
Color the tents from smallest to biggest,
then trace all the words below.

Tiny Small Mid

Big

Map The Trail!

Help the raccoon reach the campsite. Draw a line to the trail path and circle and color each forest item he finds along the way!

Bead The Bear Necklace

(PATTERN PRACTICE)

Help the bear finish each necklace pattern. Look carefully at the beads. What comes next? Color or draw the missing beads.

Pattern Type: **AB**

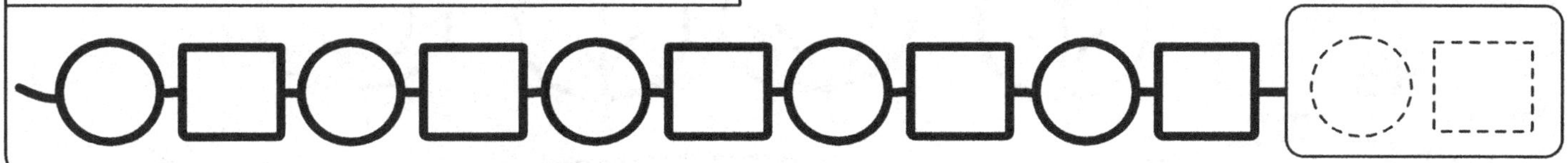

Pattern Type: **ABB**

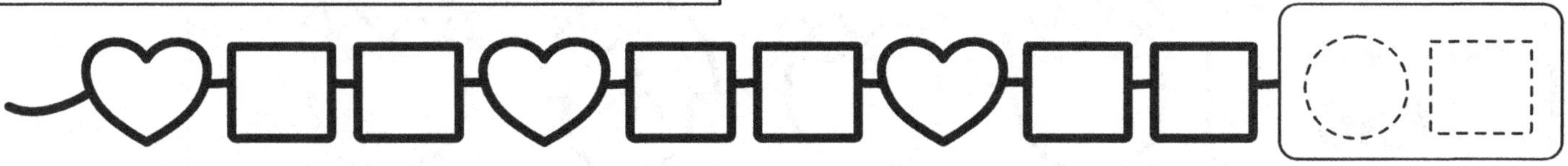

Pattern Type: **AAB**

Pattern Type: **ABC**

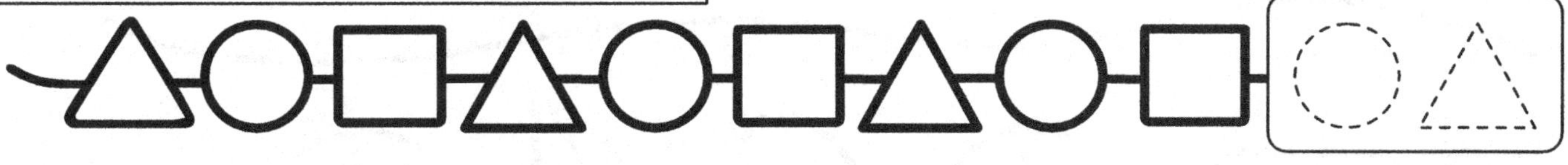

Nature Pattern Trails

Follow the trail and complete each pattern. Look at the shapes along the path and draw the next ones to finish!

Pattern Type: **AB**

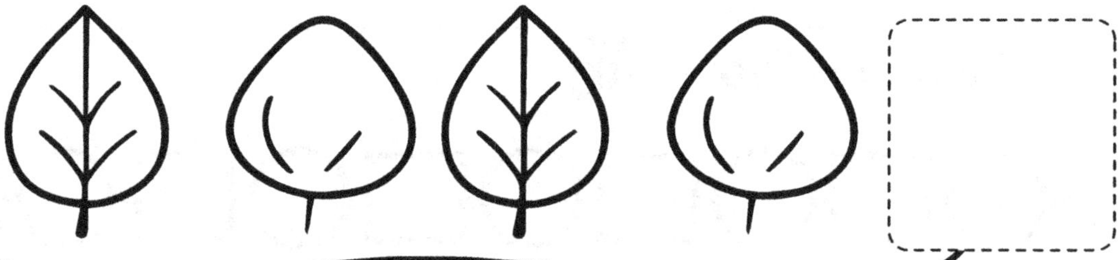

Pattern Type: **AAB**

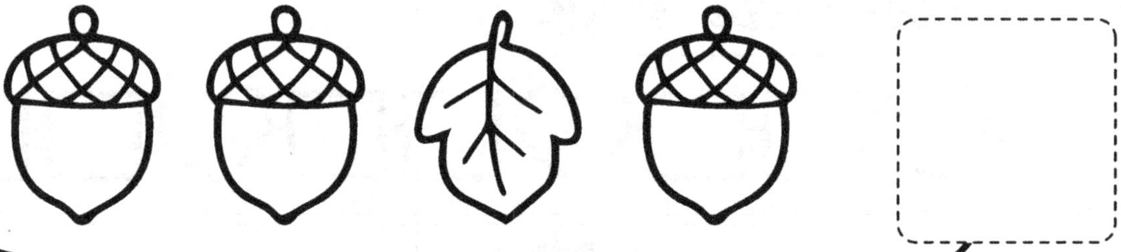

Pattern Type: **ABC**

Pattern Type: **ABC**

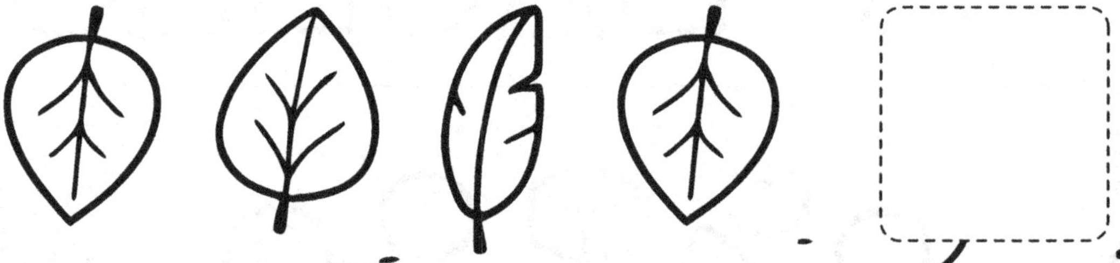

Backpack Sorting Patterns

Each camper packed a pattern!
Look inside their backpacks. What comes next?
Draw the missing gear to complete the pattern.

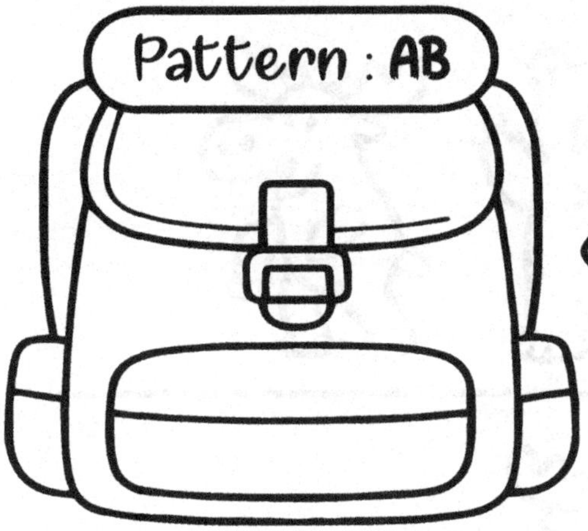

Clap – Stomp – Crawl Movement Patterns

Follow the forest friends' movement patterns!
Look at each row. What comes next? Color the next pattern

Pattern : **AB**

Pattern : **AAB**

Pattern : **ABC**

Pattern : **ABC**

Pattern Path Cut & Paste

Cut out the forest shapes at the bottom of the page.
Then look at each path above. What comes next in the pattern?
Paste the correct shape in the empty spaces to finish the trail!

→ Pattern : **AB**

Pattern : **AAB**

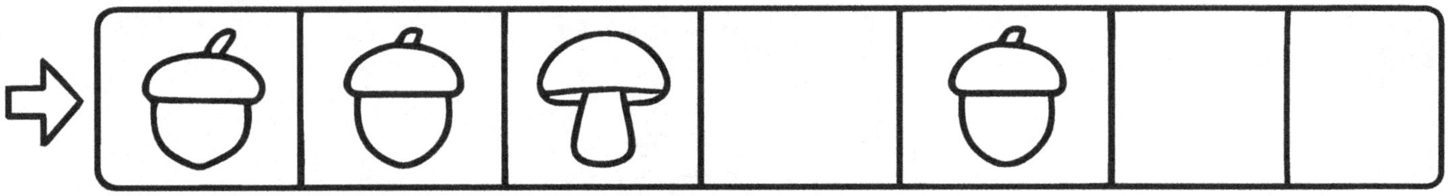

Pattern : **ABC**

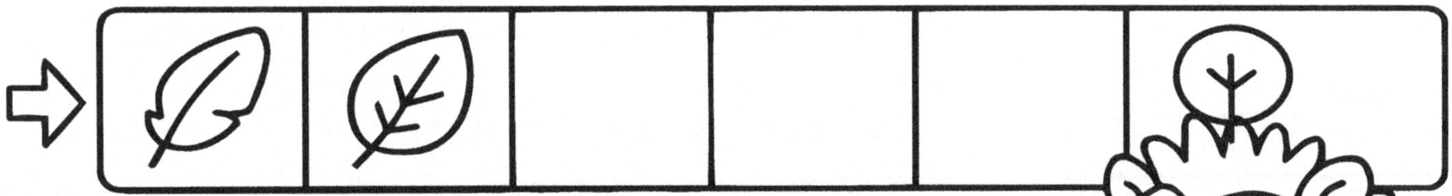

CUT OUT THESES SHAPES

WILDERNESS WARRIOR CROSS BORDER

Forest Pattern Blocks

These forest blocks follow a pattern!
Look at the shapes. What comes next?
Color the shapes to complete the trail.

Pattern : AB

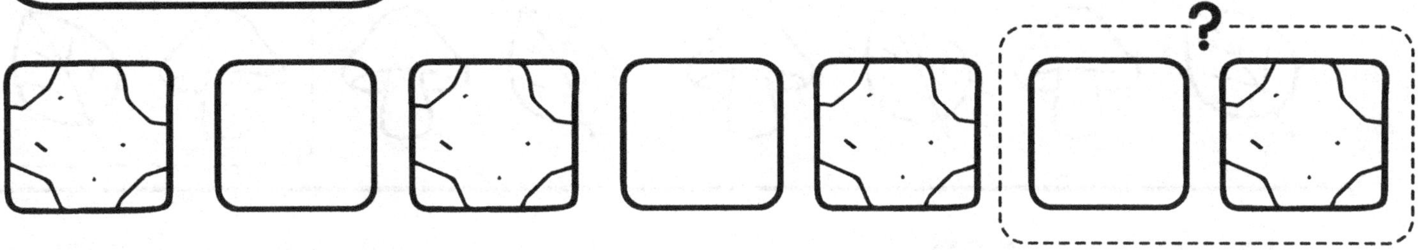

Pattern : AAB

Pattern : ABC

Finish The Pattern

Look at the pattern in each row. What comes next?
Color the missing shape to finish it!

Pattern : **AB**

Pattern : **AAB**

Pattern : **ABC**

Pattern : **ABC**

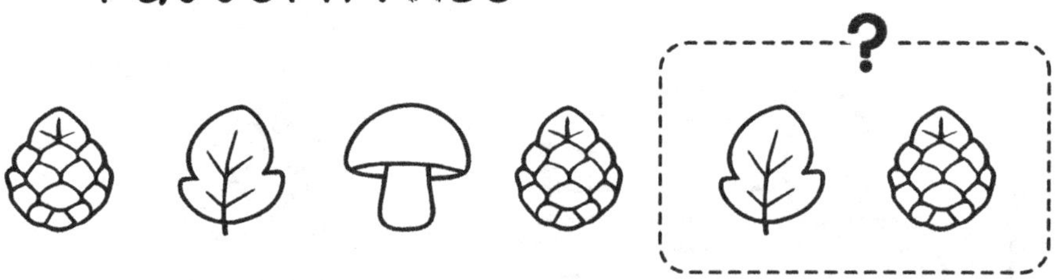

Add The Forest Friends

Count the animals in each group.
Add them together and write the answer in the box!

2 + 2 =

2 + 3 =

1 + 2 =

1 + 1 =

Snack Math
Take Away Game

Help the forest friends subtract!
Count how many snacks they had. cross out the snacks they ate. and write how many are left.

5 – 2 =

4 – 2 =

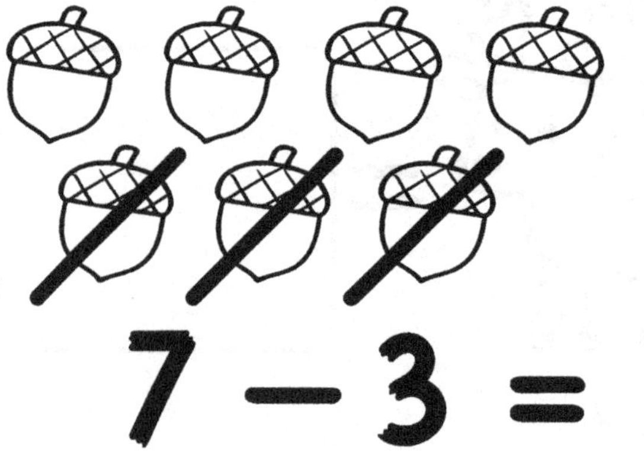

7 – 3 =

Roll & Count Dice Game

Roll a die. Count the dots.
Then color a forest friend next to that number.
Keep rolling until all the animals are colored!

1

2

3

4

5

6

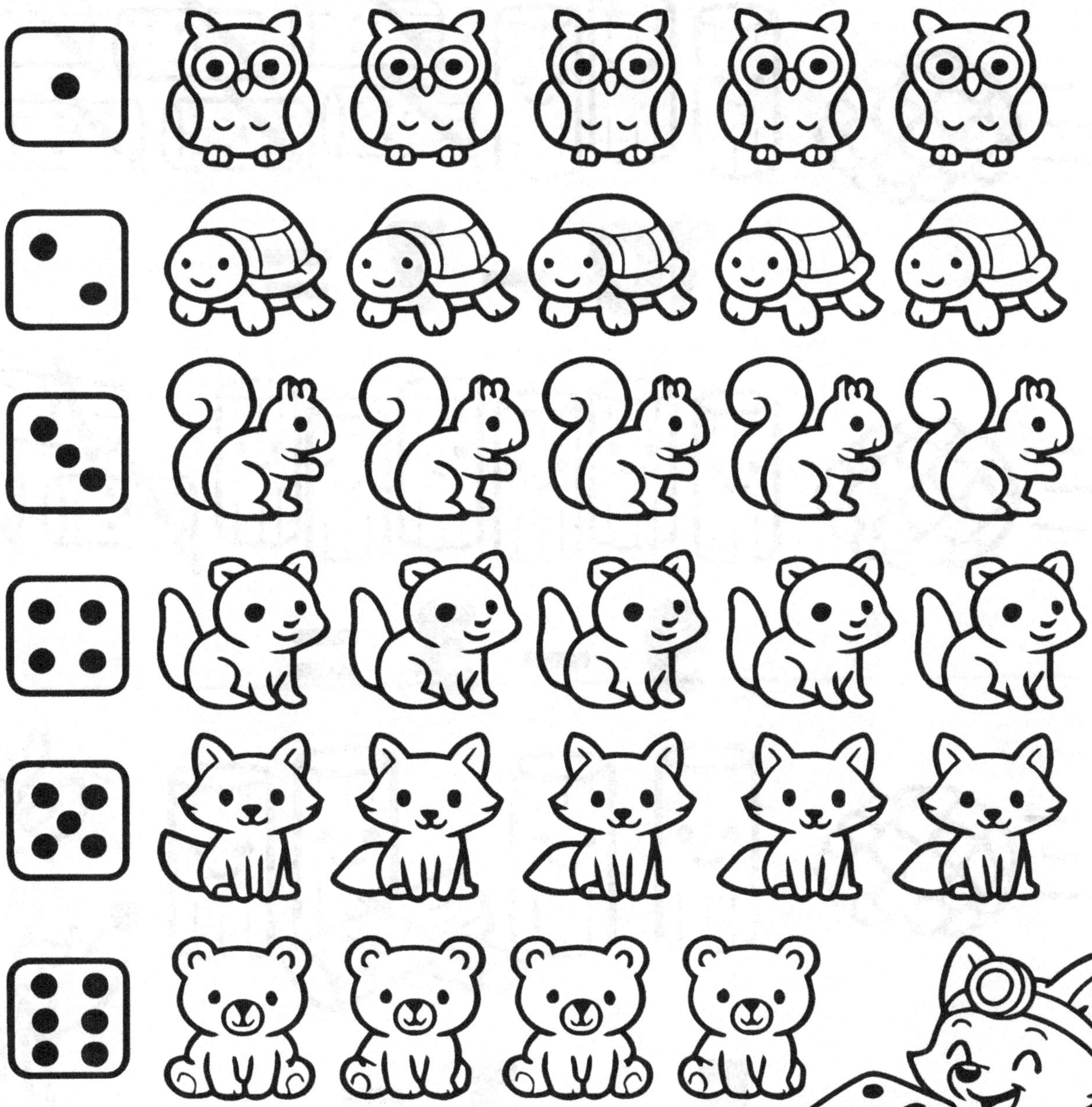

Subtraction Bowling
Pins & Pinecones

Roll the pinecone and knock down some pins!
Count how many pins stood up and how many fell.
Then write how many are left!

10 – 7 =

9 – 3 =

5 – 1 =

Add & Color Forest Word Problems

Read the word problem. Solve it.
Then color the animals to match your answer!

Two squirrels were collecting acorns.
Three more joined them.
How many squirrels are there now?

$$2 + 3 =$$

Color all the squirrel brown.

Four owls sat in a tree.
Two more flew over.
How many owls in total?

$$4 + 2 =$$

Color all the owls Grey.

Finger Song Math Fun

Use your fingers to act out each forest math rhyme.
Count along and show the answers with your hands!

3 + 1

Three sleepy owls high in a tree,
One more came -now How many are there?
Count them all, don't miss a beak,
Show with fingers, and give a squeak!

5 - 2

Five little raccoon munching on a snack,
two ran off and didn't come back!
How many raccoon are left to play?
Show with fingers -what do you say

Animal Puzzle Match

Match each forest animal to its shadow!
Draw a line to connect the shadow.

Camp Supply Sort

(PASTE)

Sort and Paste them into the correct supply bag!

Camp Supply Sort

(CUT)

Cut out the camp items below.

WILDERNESS WARRIOR CROSS BORDER

I Spy
Numbers In The Wild

Look closely at the forest scene below.
How many of each number can you find? Write below the box, count and Color them all!

Number	How many	Number	How many
0		4	
1		5	
2		6	
3		7	

Build A Rock Tower
(PASTE)

BUILD YOUR TOWER BY
PASTING ROCK CUT OUTS

Build A Rock Tower
(CUT)

Cut out the rock shapes below. Then paste them to build your own rock tower from biggest to smallest (or smallest to biggest!).

CUT OUT THE ROCKS

WILDERNESS
WARRIOR
CROSS
BORDER

Forest Maze Counting

Help the animal get through the forest maze!
Follow the path by counting in order.
Start at 1 and end at 10!

Badge Tracker

Earn a badge for each mission you complete!
Color in each badge after you finish the chapter.

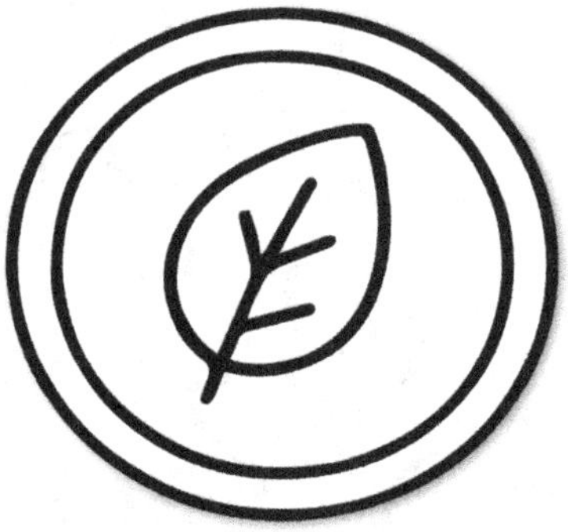

NATURE SCOUT

CAMP SAFETY PRO

WILD WEATHER WATCHER

WILDLIFE PROTECTOR

ECO HERO

CAMP CHEF

FIRE SAFETY STAR

PATTERN PRO

MATH EXPLORER

My Forest Goals

Think back on your forest adventures!
What did you do well?
What would you like to try next?
Write or draw your goals below.

MY GOAL REFLECTIONS

I feel proud of ____

____

Next time, I want to try ________________________________

DRAW YOUR GOAL!

Draw something you want to try or do better next time!

YOU'VE GOT THIS!

GET YOUR FREE BONUS PAGES!

Keep Learning With Your Wilderness Warrior Friends

Visit the Children's Books and Teacher's Corner sections for your exclusive bonus materials!

grdmpublishing.net

Need school-related help or support?
I'm here to help
teachers, parents, and schools!

Ginger@GingertheTeacher.com
www.GingertheTeacher.com

@gingertheteacher_author

GRDM Publishing, LLC

CERTIFICATE OF COMPLETION

This certifies that

has completed the
Wilderness Warrior Badge

Signature

Date

Made in the USA
Monee, IL
05 November 2026

67379291R10039